Data, Chance & Probability

Grades 6-8
Activity Book

by Graham A. Jones
& Carol A. Thornton

©1994 by Learning Resources, Inc.
Vernon Hills, Illinois 60061

All rights reserved.

This book is copyrighted. No part of this book may be reproduced, stored in a retrieval system or transmitted, in any form or by any means electronic, mechanical, photocopying, recording, or otherwise, without written permission, except for the specific permission stated below.

Each blackline master is intended for reproduction in quantities sufficient for classroom use. Permission is granted to the purchaser to reproduce each blackline master in quantities suitable for noncommercial classroom use.

ISBN 1-56911-989-9

Printed in the United States of America

Table of Contents

On Target *Teaching Notes: 5–6*
- 7 Target Toss
- 8 Line Them Up
- 9 Plot It Again
- 10 Back-to-Back

Deal the Deck *Teaching Notes: 11–12*
- 13 Black for Sure?
- 14 Equally Likely
- 15 Why Equally Likely?
- 16 Black More Likely!
- 17 Why Black?
- 18 Match the Cards
- 19 Even Things Up
- 20 The "No Hearts" Deck
- 21 Some Hearts
- 22 Which Two?
- 23 Double Red

Movie Time *Teaching Notes: 24–25*
- 26 How Many Movies?
- 27 Back-to-Back
- 28 Quartiles
- 29 Boxing the Movie Data
- 30 Movie Data Line Plot
- 31 Box Plot Scale
- 32 Favorite Movie
- 33 Plot the Rank Order
- 34 Back-to-Back Line Plot

Cups and Rollers *Teaching Notes: 35*
- 36 Roll the Roller
- 37 Cup Ups and Downs

Averages *Teaching Notes: 38–39*
- 40 Finding the Average
- 41 Zero In or Out?
- 42 Locating the Mean
- 43 Mean or Median?
- 44 Mean and Median
- 45 Add "One"
- 46 Could It Be?
- 47 What's the Total?
- 48 Mean, Median, and Mode
- 49 How Many Are Missing?

Different Ways *Teaching Notes: 50*
- 51 Pizza Today
- 52 Two from the Deck
- 53 Two Out of Four

Now Sports *Teaching Notes: 54–55*
- 56 A 50% Shooter
- 57 Two Free Throws
- 58 A 30% Average
- 59 Steffi's Serve
- 60 Up to Bat
- 61 One of Each
- 62 Stefan's Smashes

For the Record *Teaching Notes: 63*
- 64 Movie Poll
- 65 What are the Chances?
- 66 Read My Mind
- 67 Timed Experiment
- 68 Box the Timed Data

Chips *Teaching Notes: 69*
- 70 Peanut Butter Chip Cookies
- 71 No Chips!

Activity Masters
- 72 1: Target Toss
- 73 2: Spinners
- 74 2: Spinners Continued
- 75 3: Rollers
- 76 4: Cups
- 77 *Progress Chart*
- 78 *Award Certificate*
- 79 *Family-Gram*
- 80 *Good Work Certificate*

Introduction

Data, Chance, and Probability, 6-8 consists of 53 activities for students in grades 6-8. Using real-world situations and data from cross-curricular topics, students actively explore probability and data analysis.

This resource book is a carefully structured supplement to your mathematics program and can be used in regular, special education, and remedial classes. The activities enhance the development of mathematical ideas and help students understand probability and data analysis. Activities focus on:

- Data Analysis and Presentation
- Outcomes and Events
- Numerical Probabilities
- Arrangements
- Central Measures
- Mathematical Modeling

These topics follow recommendations in NCTM's *Curriculum and Evaluation Standards for School Mathematics*. Activities in this resource book address four major goals of the NCTM document: *problem solving, reasoning, communication,* and *making connections*. Activities also reflect the spirit of the NCTM 5-8 Addenda Series on *Dealing with Data and Chance*.

The problem-solving activities are presented in cooperative, active learning settings. Students:

- Explore a problem
- Predict outcomes
- Gather data and other information
- Organize the information
- Assign numbers or measurements
- Communicate results
- Reflect on and extend ideas

Each activity page presents a hands-on activity for students. Every activity has three parts:

EXPLORE:
Students first address a problem and predict the results.

TALK ABOUT IT or WRITE ABOUT IT:
With a partner, students experiment, collect and quantify data, and then discuss their results.

TELL AND SHARE:
Students review their thinking with partners, then share their findings and interpretations with the class. This sharing and discussion helps students find new applications for the ideas presented.

Most activities require that students work together with a partner. In the "Explore" part of an activity, students collect and organize data, first with a partner and then later as a group. Students are encouraged either to write or talk about their findings as a way of communicating results. Finally, in "Tell and Share," students are challenged to quantify, interpret, apply, and extend their ideas.

Teaching Notes
On Target

Warm-Up

To introduce this section, discuss ways to collect data about the distance each student lives from the public library. Direct students to write the number of blocks on a paper square, then to arrange themselves in order from those who live closest to those who live farthest. Emphasize the characteristics of the data, such as the *range* (closest to farthest), the *median* (middle distance), and the *mode* (most frequent distance).

Using the Pages

Target Toss *(page 7)*: If possible, in advance of class time prepare several Target Toss areas using masking tape. In this activity, students work with a partner to determine the total scores possible from tossing a chip at a target. Students also determine mode, median, and range.

Line Them Up *(page 8)*: On the chalkboard, draw a line and scale it from 0 to 25. Discuss the features of a line plot, including its title and a scale with equal units. Point out that data values occurring more than once are stacked on top of each other. If appropriate, note the occurrence of clusters and any gaps on the line plot. Also check to see if there is more than one mode.

Plot It Again *(page 9)*: This activity introduces a stem-and-leaf plot as a way to present data. Make sure students write large enough for their numbers to fill the page because the stem-and-leaf plot is made life-size on the floor. If necessary, point out that the stems are the tens digits of the scores, and the leaves are the ones digits.

Stem-and-Leaf Plot

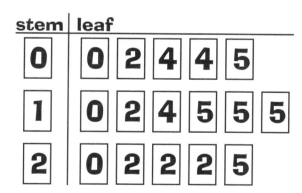

Generally, the leaves of a stem-and-leaf plot are ordered. Thus, locating the median depends on whether there is an odd or an even number of scores. When the number of scores is odd, the median is the middle score. When the number of scores is even, the median is the mean, or arithmetic average, of the two middle scores. At this level, most students can find the median in either situation.

Back-to-Back *(page 10):* This activity introduces a back-to-back stem-and-leaf plot. The two sets of data use the same stems, and the leaves are located on either side. Draw attention to the legend showing how to read the numbers on the plot.

In "Tell and Share," students are asked to tell of other situations that are appropriate for stem-and-leaf plots. These may include comparing two sporting teams on specific data, men's and women's earnings in tennis or golf, the television ratings of two hits over a 12-week period, or the cost of renting videos this year and last year.

Wrap-Up

Encourage students to find examples of data in their textbooks, newspapers, or sporting magazines that could be organized using line plots or stem-and-leaf plots. Have students display their data and their corresponding plots.

Target Toss

Name_____

 **Explore:**
- Set up the target.
- Stand 5 paper lengths (or 5 feet) away.
- Toss the chip 5 times.
- Record your total score (0 if you miss).

Data Activity

- Find a partner.
- List the total possible scores from low to high.
- Where did each of your scores fall on the list?

Use the words in the Word Box to describe your scores.

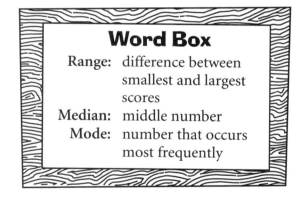

Word Box
Range: difference between smallest and largest scores
Median: middle number
Mode: number that occurs most frequently

Was your list the same as others in the class?

Need: 1 red-white chip, paper, and Activity Master 1.
Save: *Target Toss* total score for pages 8 and 9.

Line Them Up

Name _____

 **plore:** • Write your total score in *Target Toss* on the back of the paper square.

Class Line Plot

- Work as a class.
- Make a large line plot on the chalkboard as shown here.
- Tape your ⊠ above the appropriate point on the chalkboard scale.

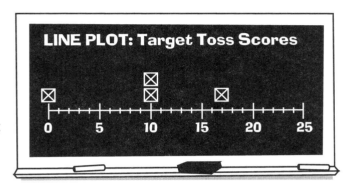

 Use the words in the Word Box to describe your scores.

Word Box
Range: difference between smallest and largest scores
Mode: number that occurs most frequently

 Are there gaps on the line plot?

Why?

Need: For each student, a 2" marked paper square and score from *Target Toss,* page 7.

8

Data, Chance and Probability Activity Book, 6–8
© 1994 Learning Resources, Inc.

Plot It Again

Name_____

 **Explore:**
- Write your total score in *Target Toss* large on paper.
- Look at your score.
- If your score has 2 digits, fold the tens digit under.

Stem-and-Leaf Plot

- Work as a class.
- Make a stem-and-leaf plot on the floor as shown at the right.
- Look at the tens "stems" to decide on your row.
- Place your paper beside your stem.
- Work with other "leaves" in your row to order the ones from low to high.

stem	leaf
0	0 2 4 4 5
1	0 2 4 5 5 5 7
2	0 2 2 2 5

Use the words in the Word Box to describe your scores.

Word Box
- **Range:** difference between smallest and largest scores
- **Median:** middle number
- **Mode:** number that occurs most frequently

- How are the stem-and-leaf plot and the line plot alike?
- How are they different?

Need: 8½" × 11" paper, tape to make the plot, and score from *Target Toss*, page 7.
Save: *Plot it Again* scores for page 10.

Back-to-Back

Name _____

Explore:
- Your teacher will divide the class into two groups.
- Use your paper from *Plot it Again*.
- Find the range, median, and mode for each group.

Stem-and-Leaf Plots

- Determine your row.
- Place your paper beside your stem.
- Work with others in your row to order the "leaves" from low to high.

Back-to-Back Stem-and-Leaf Plot

Group A		Group B
5 4 4 4 2	0	0 2 2 4 5
7 7 5 4 4 2	1	0 2 4 5 5 7 9
5 0	2	0

1|5 = 15 points

How do the two groups compare?

In what other situations could you use a back-to-back stem-and-leaf plot?

Need: Student scores from *Plot It Again,* page 9.

Teaching Notes
Deal the Deck

Warm-Up

These activities lead to the assignment of numerical probabilities. Have students concentrate on certain and impossible events to set the limits for probability, namely, 1 (certain) and 0 (impossible). Encourage students to identify "certain" and "impossible" events throughout the school day, and assign 1 or 0 to each event. As an extension, have them identify some possible (but not certain) events, and predict their probabilities between the extremes of 1 and 0.

Using the Pages

Black for Sure? *(page 13):* Students focus on certain, possible, and impossible events, and assign numerical probabilities of 1 for *certain* events and 0 for *impossible* events. This activity sets the stage for determining all numerical probabilities. In "Tell and Share" the probability of drawing a red from Deck 1 is 1/3; the probability of drawing a red from Deck 2 is 0.

Equally Likely *(page 14):* Students concentrate on events that are equally likely, or have the same probability of occurring. Encourage them to use equivalent expressions for equally likely such as "same chance," "equal probabilities," or "fair for all outcomes."

Why Equally Likely? *(page 15):* Students combine the tallies from *Equally Likely* on page 14 in a class graph. In "Tell and Share," students should show the chances of drawing each color is 1/2, since the outcomes are equally likely.

Black More Likely! *(page 16):* With the deck of cards, students have twice as many black cards as red cards. The results of the experiment should show that a black card will be drawn about 2 out of every 3 times.

Why Black? *(page 17):* Emphasize to students that combining their results into class data provides a more stable measure of probability.

Match the Cards *(page 18):* This activity is a simple model. Stress that the model matches the outcomes as well as their probabilities. This correlation will be important in more complex simulations.

Even Things Up *(page 19):* In this activity there are several possible ways to make red and black equally likely. One way is to remove 13 black cards. Another way is to add 13 red cards. In "Tell and Share," the responses will vary.

You might observe that extreme results occur in the experiment because of the small number of trials. To obtain a more "average" or precise picture of what happens, you can combine the data and compare results.

Data, Chance and Probability Activity Book, 6–8
© 1994 Learning Resources, Inc.

The Packer's Deck and **Some Hearts** *(pages 20 and 21):* Be sure students understand the terms *suits* and *face cards* and know the numbers in each category.

Which Two? *(page 22):* Since the cards are drawn *with replacement,* the probability of drawing two red cards is 1/4, based on the tree diagram as shown. Note that in each branch, the probability of drawing a red card is 1/2, since the card is replaced after the first drawing. As a result, any estimate in the range of 1/4 is a reasonable experimental probability.

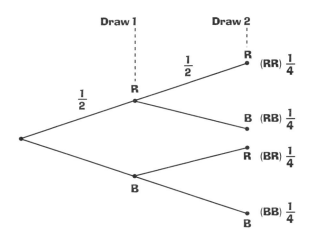

Double Red *(page 23):* The combined class data should give results reasonably close to 1/4.

 ## Wrap-Up

Ask students to identify two situations: one involving coins in which the events are equally likely, and another in which one event is twice as likely to occur.

Black for Sure?

Name _____

 **plore:** What color will be drawn first from each deck if:
- Deck 1 has an equal number of clubs, spades, and diamonds.
- Deck 2 has only spades and clubs.

Your Turn
- Find a partner.
- Make Deck 1 and Deck 2.

The Experiment
- Use Deck 1.
- Draw one card, and tally the color.
- Replace the card and shuffle.
- Repeat 5 times.
- Partner takes a turn.

Tally Sheet
Deck 1

Red	Black

Why is it unnecessary to tally the color of the cards drawn from Deck 2?

Use terms from the Word Box to compare the chances of drawing a black card from the two decks.

Word Box
- **Certain:** an event that will always occur
- **Possible:** result may occur
- **Impossible:** an event that will never occur

- On the scale, show the chance of drawing a red card for each deck.
- For Deck 1, mark R1. For Deck 2, mark R2.

0 — Impossible ———————— 1 — Certain

Need: 2 decks of cards.

Data, Chance and Probability Activity Book, 6–8
© 1994 Learning Resources, Inc.

Equally Likely

Name_____

 **Explore:** Which color will be drawn first?   26 ♡ ◇
                                                                  26 ♠ ♣

Your Turn

- Find a partner.
- Make the deck of cards shown in "Explore."

The Experiment

- Shuffle the cards.
- Draw one card, tally the color, and replace the card.
- Repeat 15 times.
- Partner takes a turn.

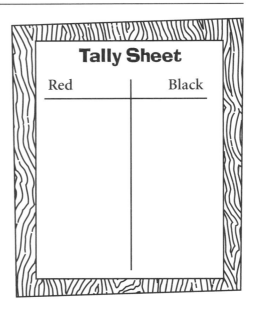

Tally Sheet

Red	Black

Did the experiment turn out as you predicted?

How do your results compare with those of your classmates?

Was any color drawn more often?

Did this surprise you? Why?

Need: Deck of cards. **Save:** *Equally Likely* Tally Sheet for page 15.

Why Equally Likely?

Name_____

 **plore:** In a regular card deck, is the chance of drawing a red card first the same as for a black card?

Graph Activity

- Work as a class.
- Combine the group totals from *Equally Likely*.
- Use tally blocks of 5 to make the class graph.
- Total the number for:

  Black_____   Red_____

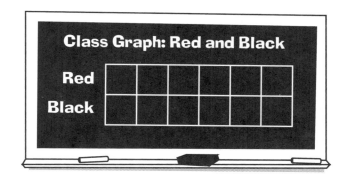

Compare red and black on the class graph.

What does this say about the chances of drawing a red card first?

A black card first?

On the scale, show the chance of drawing each color first:
- red (R)
- black (B)

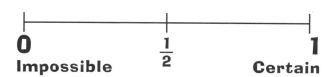

Need: Data from *Equally Likely*, page 14.

Data, Chance and Probability Activity Book, 6–8
© 1994 Learning Resources, Inc.

Black More Likely!

Name_____

 **Explore:** From this "no hearts" deck, which color will be drawn first?

13 of each

Your Turn

- Find a partner.
- Make the deck of cards as shown above.

The Experiment

- Shuffle the cards.
- Draw one card, tally the color, and replace the card.
- Repeat 5 times.
- Partner takes a turn.

Did the experiment turn out as you predicted?

What fraction of the time was the card red?

What fraction of the time was the card black?

Need: Deck of cards. **Save:** *Black More Likely!* Tally Sheet for pages 17 and 18.

Why Black?

Name_____

 **Explore:** In the "no hearts" deck, is the black card more likely than a red card to be drawn first?

Graph Activity

- Work as a class.
- Combine the group totals from *Black More Likely*.
- Use tally blocks of 5 to make the class graph.
- Total the number for:

  Red_____ Black_____

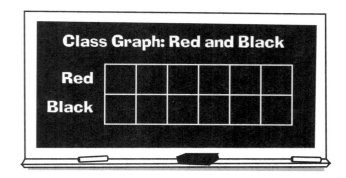

 What fraction of the total number of tally marks are black?

What does this say about the chances of drawing black first?

 On the scale, show the chance of drawing each color first:
- red (R)
- black (B)

What is the sum of the chances?

Why?

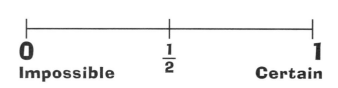

Need: Data from *Black More Likely*, page 16.

Data, Chance and Probability Activity Book, 6–8
© 1994 Learning Resources, Inc.

Match the Cards

Name _____

 **plore:** Which spinners work like the cards in the deck in *Black More Likely*?

- Label each part B(black) or R(red).

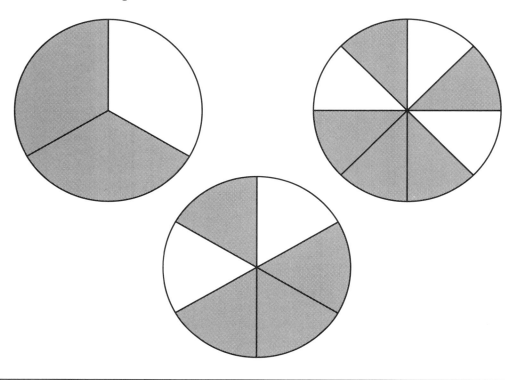

Tell why the spinners you chose are like the cards in the deck.

Can you make another spinner that would work? Show it.

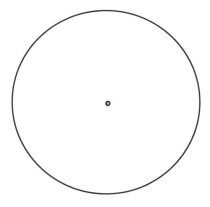

Need: Data from *Black More Likely*, page 16.

18

Data, Chance and Probability Activity Book, 6–8
© 1994 Learning Resources, Inc.

Even Things Up

Name _____

 **Explore:** How could you change the deck to make the chance of drawing a red or a black card the same?

13 of each

Your Turn

- Find a partner.
- Change the deck so the chance of drawing a red or black card is the same.

The Experiment

- Shuffle the cards.
- Draw a card, tally the color, and replace the card.
- Repeat 5 times.
- Partner takes a turn.

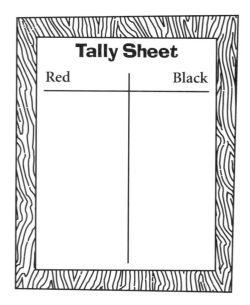

Did the experiment turn out as you predicted?

How do your results compare with those of other classmates?

If you found 5 extra red cards, what could you do to the original deck to make the chance of drawing a red or black card equally likely?

Need: Deck of cards.

Data, Chance and Probability Activity Book, 6–8
© 1994 Learning Resources, Inc.

The "No Hearts" Deck

Name _____

 **plore:** From this "no hearts" deck of cards, which suit do you think will be drawn first?

13 of each
♢ ♠ ♣

Your Turn
- Find a partner.
- Make the "no hearts" deck of cards.

The Experiment
- Shuffle the cards.
- Draw a card, tally the suit, and replace the card.
- Repeat 10 times.
- Partner takes a turn.

Did the experiment turn out as you predicted?

How do your results compare with those of other classmates?

What fraction of the time was the card a ♢? ♠? ♣?

On the scale, show the chance of drawing each suit first:
- diamond (D)
- club (C)
- spade (S)
- heart (H)

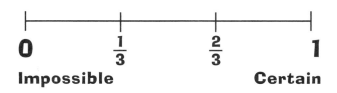

Need: Deck of cards.

Some Hearts

Name _____

 **Explore:** This deck has some hearts missing. How could you change the deck so the chance of drawing each suit is equally likely?

13 of each
◇ ♠ ♣
and 4 ♡'s

Your Turn

- Find a partner.
- Change the deck so the chance of drawing each suit is equally likely.

The Experiment

- Shuffle the cards.
- Draw a card, tally the suit, and replace the card.
- Repeat 5 times.
- Partner takes a turn.

Did the experiment turn out as you predicted?

Suppose you repeated the experiment 200 times. Would the outcome be the same? Explain.

If you found 4 extra hearts, what could you do to the original deck to make the chance of drawing any suit equally likely?

Need: Deck of cards.

Which Two?

Name _____

 **plore:**
- Draw a card.
- Replace it, shuffle and draw another card.
- Mark an X on the scale to show the chance of drawing 2 red cards.

26 red
26 black

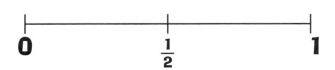

Your Turn

- Find a partner.
- Experiment with the deck of cards.

The Experiment

- Shuffle the cards.
- Draw one card, note its color and replace it.
- Shuffle again, and draw another card.
- Tally the result and repeat 10 times.
- Partner takes a turn.

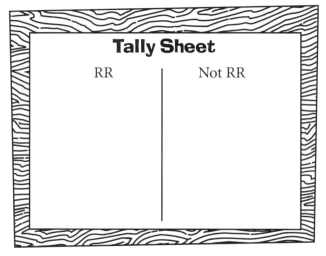

Tally Sheet

RR	Not RR

 What fraction of the tallies were RR?

 How does the fraction for RR compare with your estimate?

Need: Deck of cards. **Save:** *Which Two?* Tally Sheet for page 23.

Double Red

Name_____

 **Explore:** Is the chance of drawing 2 cards in *Which Two* greater than or less than 50%?

Graph Activity

- Work as a class.
- Combine the group totals from *Which Two*.
- Use 1 square for each tally block of 5.
- Tape the squares to make a class graph.
- Total the number for:

  RR_____ not RR_____.

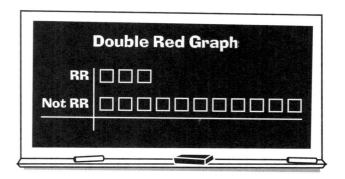

How many squares are there in all?

About what fraction of these are RR?

What does this say about the chances that someone will draw two red cards?

On the scale, mark an X to show the chance of drawing two red cards.

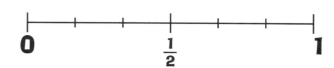

Need: Data from *Which Two*, page 22, and 2" X 2" paper squares.

Teaching Notes
Movie Time

Warm-Up

This section focuses on collecting and presenting data. Students will present data in three different formats: box-and-whisker plots, line plots, and stem-and-leaf plots. Before students begin the activities, ask them to do the following:

- list all the movies they have seen during the last two weeks;
- list the best movies they have seen this year.

Using the Pages

How Many Movies? *(page 26):* Since students have had some experience making stem-and-leaf plots, expect them to take more initiative in building the plot. If necessary, review the procedure.

At this level, do not expect a precise measure of an outlier. There is a specific way to determine outliers for a set of data. Upper outliers lie 1.5 times the interquartile range above the upper quartile. Similarly, lower outliers lie 1.5 times the interquartile range below the lower quartile. Note that the interquartile range is the difference between the upper and the lower quartiles.

Back-to-Back *(page 27):* Students use the same data as the previous activity, but the class is divided into two groups.

Quartiles *(page 28):* In "Talk About It," 25% of the data are below the lower quartile. Similarly, 50% of the data are below the median, and 75% of the data are below the upper quartile. These percentages are important for box plots because the two whiskers each contain 25% of the data, and the box contains 50% of the data, with 25% on each side of the median.

In "Tell and Share," if one more student joined the class, it is possible but not certain that both the median and the upper quartile would change. For example, if there are 3 pieces of data, that is, 1, 2, 3, the median is 2. If 5 is added, the median becomes 2.5. On the other hand, if the original data are 1, 2, 2 (median 2), the addition of a fourth piece of data such as 5 would not change the median. Similar examples can be constructed for upper quartile values.

Boxing the Movie Data *(page 29):* Note that this activity is an informal introduction to box-and-whisker plots. There is no accompanying scale to the plot, and data points are merely rank ordered. A scale is included in the activity, *Box Plot Scale,* page 31.

Movie Data Line Plot *(page 30):* After the line plot is completed, keep it on the board for the next activity, *Box Plot Scale*.

Box Plot Scale *(page 31):* This activity introduces a number line as the scale for a box-and-whisker plot. In "Tell and Share," 50% of the data lie within the box, and 50% of the data lie outside the box, on the two whiskers.

Favorite Movie *(page 32):* Students in the class make a list of their four favorite movies. Next, each student ranks these movies: 4 (favorite) to 1 (least favorite). Then the total score of all students is found for these movies.

Plot the Rank Order *(page 33):* Students compare the highest rank order total score and the movie that received the greatest number of number 4 rankings (favorite).

Back-to-Back Line Plot *(page 34):* In this back-to-back line plot, students compare the number of each weighted choice for the two movies with the top total scores.

Wrap-Up

Ask students to choose another "Favorite_____." Students will need to decide on a way of rating the data and plotting it so comparisons can be made.

How Many Movies?

Name _____

 **Explore:** About how many movies did you and your classmates watch during the last two weeks?

Data Collection

- Make a list of the movies you have watched during the last 2 weeks.
- Write the number on a paper square. If the number has 2 digits, fold the tens digit under.
- Work as a class. Make a stem-and-leaf plot on the chalkboard.
- Add your paper to the plot.
- Let the first student put the legend on the plot.
- Let the last 2 students order the leaves.

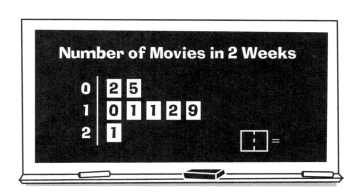

 Use the words in the Word Box to describe the class plot.

Word Box
Range: difference between smallest and largest numbers
Median: middle number
Mode: number that occurs most frequently
Outlier: extreme value

 What does this stem-and-leaf plot tell you about the number of movies watched?

Need: 1 small paper square and tape. **Save:** *How Many Movies?* for pages 27 and 30.

Back-to-Back

Name_____

 **Ex)plore:**
- Divide into 2 groups.
- Use your paper from *How Many Movies*?

Data Collection

- Determine your group.
- Place your paper square beside your "stem."
- Work with others in your row to order the "leaves" from low to high.

 Use the words in the Word Box to describe the plot.

Word Box

Range: difference between smallest and largest numbers
Median: middle number
Mode: number that occurs most frequently
Outlier: extreme value

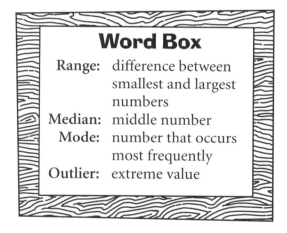

 What other situations could you use to compare data in this way?

Need: Paper squares from *How Many Movies*, page 26.

Quartiles

Name _____

 **Explore:** How does the number of movies you watched during the last 2 weeks compare with other classmates?

Class Activity

- Work as a class.
- Use your paper square from *How Many Movies?*
- Arrange the numbers on the chalkboard from lowest to highest number of movies.
- Find the median.
- Find the upper quartile.
- Find the lower quartile.

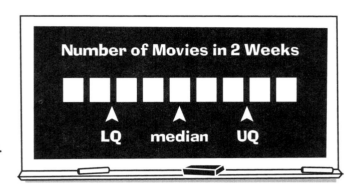

 What fraction of the data is below:

- The lower quartile?

- The median?

- The upper quartile?

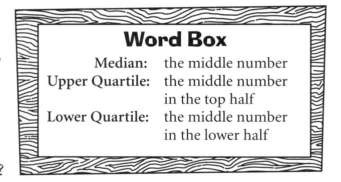

Word Box
Median: the middle number
Upper Quartile: the middle number in the top half
Lower Quartile: the middle number in the lower half

 If one more student joined the class, would the median change?

Would the upper quartile change?

Need: Paper squares from *How Many Movies*, page 26.
Save: Paper squares for *Boxing the Movie Data*, page 29.

Boxing the Movie Data

Name _____

 **Explore:** What was the largest number of movies watched by a student? What was the smallest number?

Class Activity

- Work as a class at the chalkboard.
- Use the paper squares from *Quartiles*.
- Draw the vertical median line.
- Draw the vertical lines for the "upper" and "lower" quartiles.
- Complete the box.
- Place a dot at the lowest and the highest numbers.
- Complete the whiskers.

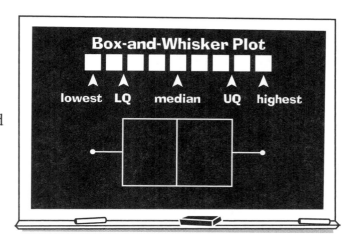

 What does the box-and-whisker plot show?

 How many movies did the typical student watch in 2 weeks?

What can be said about the movie habits of the typical student?

Need: Paper squares from *Quartiles,* page 28.

Movie Data Line Plot

Name_____

 **Explore:** Mark the scale below to show how many movies you watched during the last 2 weeks.

Class Activity

- Work as a class to make a line plot.
- Draw a large scale on the chalkboard.
- Draw a square above the scale for each classmate.
- Find the median. Color it red.
- Find the upper and lower quartiles. Color them blue.
- Color the squares for the lowest and the highest numbers yellow.

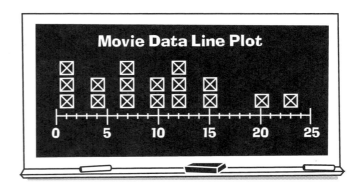

 How does the scale help you interpret the data?

 How is the line plot like the box-and-whiskers plot?

How is it different?

Need: *How Many Movies,* page 26; red, blue, and yellow chalk.
Save: *Movie Data Line Plot* for page 31.

Box Plot Scale

Name_____

 **Explore:** Where do you fit on the box-and-whisker plot on page 29—in a box or on a whisker?

Class Activity

- Work as a class.
- Use the line plot from *Movie Data Line Plot* to construct a box-and-whisker plot.
- Mark the lowest and highest points.
- Draw in the vertical lower quartile (LQ) and upper quartile (UQ) lines.
- Draw in the vertical median (M) line, and complete the box-and-whisker plot.

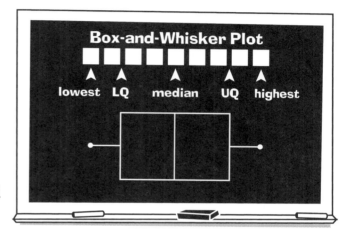

 What fraction of the numbers lie within the box?

What fraction of the numbers lie in the upper whisker? Lower whisker?

 Compare the percentage of students with numbers inside the box with the percentage outside the box. How do the percentages relate to each other?

Need: *Movie Data Line Plot*, page 30.

Favorite Movie

Name_____

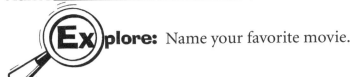 **Explore:** Name your favorite movie.

_____

Class Activity

- Work with the class.
- Make a list of everyone's favorite movies.
- List the top 4 movies.
- Rank each of these four movies 1, 2, 3, or 4. (4=most liked, 1=least liked)
- Find the class total for each movie.

Tally Sheet

Movie	Rank: 4	3	2	1	Total Score
A					
B					
C					
D					

 What do the class totals tell you about the rank order of the movies?

 Would a different method have produced different results for the rank order?

Save: *Favorite Movie* for pages 33 and 34.

Plot the Rank Order

Name_____

 **Explore:** How could the scores of the 4 movies be displayed?

Class Activity

- Work as a class.
- Choose a scale. Start with 0.
- Plot each movie's total score.

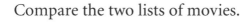
0

List the movies in order based on their total scores:

1

2

3

4

Now list the movies in order based on the number of "most liked" votes:

1

2

3

4

Compare the two lists of movies.

Was the same movie first in each list?

Need: *Favorite Movie*, page 32. **Save:** *Plot the Rank Order* for page 34.

Back-to-Back Line Plot

Name _____

 **Explore:** How could all the data from the top two preferred movies be displayed?

Class Activity

- Work as a class.
- Make a back-to-back line plot.
- Each student puts an X beside the rating given to each of the top two movies.

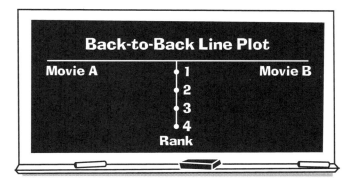

 Describe the display.

 Compare this line plot and the line plot of the total scores from page 33.

Need: *Favorite Movie*, page 32, and *Plot the Rank Order*, page 33.

Teaching Notes
Cups and Rollers

Warm-Up

To determine the chance of an event happening, it is not always possible to base the result on geometry or symmetry. Try the following demonstration: Take a nickel, place it on its edge on the overhead or a table, where it can be seen by all. Ask students to predict the probability of "heads" or "tails" when the surface is tapped hard enough to make the coin fall. Repeat at least 10 times. (The probability should be close to 0.9 for "heads.") Ask students to examine the nickel closely and explain why this is the case.

Using the Pages

Roll the Roller *(page 36)*: In this activity students consider three outcomes: two ends and a side. To simplify the experiment, have them consider only two outcomes: "on the side" or "on the end." In "Tell and Share," students should conclude that the probability of falling "on the side" is greater for Roller 2 than for Roller 1.

Cup Ups and Downs *(page 37)*: In "Tell and Share," students should realize that the probability of a cup with a pinhole-size top falling "on the top" is almost 0.

Wrap-Up

Challenge students to make other cups and rollers that have extreme probabilities of falling in particular ways.

Data, Chance and Probability Activity Book, 6–8
© 1994 Learning Resources, Inc.

Roll the Roller

Name_____

 **plore:** How many ways can the roller fall? Predict which outcome will occur more often. On the scale, show the probability for this outcome with an X.

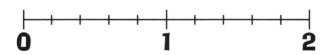

Your Turn

- Find a partner.
- Make Roller 1 and Roller 2.
- Experiment with Roller 1.

The Experiment

- Drop the roller from waist level.
- Tally the outcome.
- Repeat 10 times.
- Partner takes a turn.

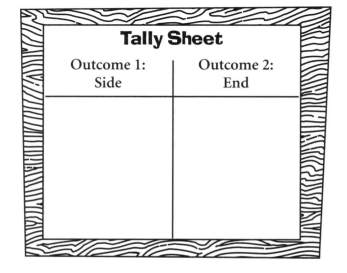

Tally Sheet

Outcome 1: Side	Outcome 2: End

- Did the experiment turn out as you predicted?
- How do your results compare with those of other classmates?
- Add your tally to the class tally.
- Based on the class tally, what is the probability of each outcome?

Would the probabilities of the outcomes be the same for Roller 2? Explain your reasoning.

Need: Activity Master 3, tape, and scissors.

Cup Ups and Downs

Name_____

 **Explore:** How many ways can the cup fall?
- Predict which outcome will occur the least.
- On the scale, show the probability of this outcome with an X.

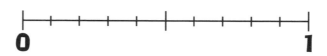

Your Turn
- Find a partner.
- Make Cup 1 and Cup 2.
- Experiment with Cup 1.

The Experiment
- Drop the cup from waist level.
- Tally the outcome.
- Repeat 10 times.
- Partner takes a turn.
- Add your tally to the class tally.

Tally Sheet

Outcome 1: Small End	Outcome 2: Large End	Outcome 3: Side

 Based on the class tally, what is the probability of each outcome?

What do you notice about the sum of the probabilities?

 Do you think the probabilities are the same for Cup 2? Explain.

How could you make the probability of an outcome be near 0?

Need: Activity Master 4, tape, and scissors.

Teaching Notes
Averages

Warm-Up

Tell students to write down the time they went to bed last night. Tally the times to the nearest half hour. Ask students to suggest a measure that reflects the typical bedtime. Since students already have considered the mode and median in earlier activities, they are likely to volunteer these approaches. Discuss the concept of *arithmetic mean.* Next, use the three measures to identify "typical" students of the group. Evaluate which measure is best for the bedtime data.

The activities in this section focus on the arithmetic mean, or simply the *mean.* Assign students to work in groups of 4, except on pages 43 and 48. Most activities in this section follow these steps:

- Counters are distributed unequally among the four students.
- Students record the total number of counters, the number of counters each received, and the number of students in the group.
- Students then redistribute the counters so each person has the same number.
- The new number is identified as the *mean*.

Using the Pages

Finding the Average *(page 40):* This activity helps to develop an intuitive understanding of means and how they are calculated. In "Write About It," students need to realize that the mean is between the smallest and largest number of counters students receive in the group. They also should begin to realize that the mean is found by dividing the total number of counters by the number of people receiving them.

Zero In or Out? *(page 41):* Zero is used to find the mean. For example, in "Tell and Share," find the mean by dividing (4 + 0 + 2 + 6) by 4.

Locating the Mean *(page 42):* In this activity, students should find that the mean lies between the least and greatest numbers.

Mean or Median? *(page 43):* Divide the class into four groups. The mean is affected by an extremely high or low score, while the median and the mode usually are not affected by extreme scores. Students explore this concept by examining two sets of data that differ only in the least number of counters. In each case, the median is 32. However, the mean shifts from 32 to 25 when the lowest number of counters is changed from 30 to 2. In "Tell and Share," the following two sets of data shift the mean from 2 to 5: 1, 2, 12 and 10, 2, 3.

Mean and Median *(page 44):* Students work in groups of four. This activity examines the effect that extreme scores have on the mean and the median. In "Tell and Share," possible responses are: 1, 2, 2, 3; 1, 3, 3, 5; or 3, 3, 3, 3.

Add "One" *(page 45):* In "Explore," some students may find the answer either by mentally redistributing the counters or by calculating the mean. Students should realize that if the number of counters added is *greater* than the mean, then the mean *increases*; if the new number is *less* than the mean, then the mean *decreases*. If the new number is *equal* to the mean, the mean is *unchanged*.

Could It Be? *(page 46):* Students discover that the mean may not be a number in the data set. They also see that fractions and decimals are needed to express the mean of a data set that contains only whole numbers. In "Explore," the data are 2, 2, 3, 3, but the mean is 2.5. In "Tell and Share," students should determine that the only possible number is 4.

What's the Total? *(page 47):* Each group of four students finds the total number of counters, given the mean.

Mean, Median, and Mode: *(page 48):* Divide the class into groups of 5. This activity contrasts the mean, median, and mode. The "Explore" problem has several correct data sets. Two possibilities are 2, 4, 4, 4, 6 and 1, 3, 4, 4, 8. This problem provides an opportunity for students to explore, reason, and communicate their ideas.

How Many are Missing? *(page 49):* Students work again in groups of 4. The "Explore" section presents a challenging problem-solving experience. One way of reasoning is as follows: "The total for three students is 12, since the mean is 4. So the total for four students is 20, since the mean is 5. Therefore, the fourth student needs 8 counters. In "Tell and Share," there will be five students, so the total is 25 when the mean is 5. The two new students should bring 13 counters in all.

Wrap-Up

Provide each group of 4 students with an envelope containing digit cards for 0–9. Ask students to model and solve the following sample question, then explain their reasoning to the class.

Sample Question: Three of four students in a group have 1, 1, and 3 counters. If the fourth student draws a number from the envelope, how will this affect the mean? The median? The mode?

Finding the Average

Name _____

 **Explore:** Each group member should pick some counters from the cup so that all 12 counters are used.

How many counters did you get?

Your Turn

- Record the number of people in your group.
- Record the number of counters for each person in your group.
- Share so that each person has the same number of counters.

How many counters does each person have now? This is the average, or mean: _____

 Describe the number 3 in relation to the total number of counters and the number of counters first taken by each student.

Word Box

Mean: the shared number for the group

 If the total number of counters is 16, what is the mean?

Need: 12 counters for each group of 4 students.

Zero In or Out?

Name _____

 **plore:** If each group member gets the number of counters shown, estimate the mean: _____

    1 7 4 0

Your Turn

- Record the number of people in your group.
- Each person takes counters to match a number shown above.
- Record the number of counters for each person and the total for the group.
- Share so that each person has the same number of counters.

How many counters does each person have now? This is the mean: _____

 Describe how you found the mean.

Did you use the 0?

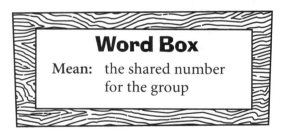

Word Box

Mean: the shared number for the group

 Suppose the numbers are 4, 0, 2, 6. What is the mean?

Need: 12 counters for each group of 4 students.

Locating the Mean

Name_____

plore: If the group gets the number of counters shown below, between which two numbers does the mean lie?
_____ and _____.

| 6 | 10 | 8 | 8 |

Your Turn

- Record the number of people in your group.
- Each person takes counters to match a number shown in "Explore."
- Record the number of counters for each person and the total for the group.
- Share so that each person has the same number of counters.

How many counters does each person have now? This is the mean: _____

Was the mean between the numbers you predicted?

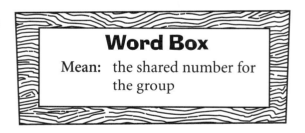

Word Box

Mean: the shared number for the group

How would you describe these two numbers?

Suppose the recorded numbers are 2, 10, 10, 10. Between which two numbers does the mean lie?

Need: 32 counters for each group of 4 students.

Mean or Median?

Name_____

 **Explore:** There are 4 tubs of counters as shown. What is the mean number of counters in a tub?

The median number?

Class Activity

- Record the number of groups in your class.
- Each group takes a tub with 34, 32, 32, or 2 counters.
- Work as a class.
- What is the median number of counters?
- Share so that each group has the mean number of counters. What is the mean?

How have the tubs of counters changed from "Explore" to the "Class Activity?"

How did this affect the mean? The median?

Change one number so the mean shifts from 2 to 5. Did the median change?

Need: 4 groups of 32 counters for the class.

Data, Chance and Probability Activity Book, 6–8
© 1994 Learning Resources, Inc.

Mean and Median

Name_____

 • If each group member gets one of the number of counters shown below, find the median and the mean.

Your Turn

- Record the number of people in your group.
- Each person takes counters to match a number shown in "Explore."
- Record the number of counters for each person and the total for the group.
- Share so that each person has the same number of counters.

How many counters does each person have now? This is the mean.

If necessary, order the original number of counters each person had to find the median.

 Compare the median and the mean.

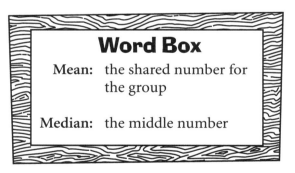

 Why are the median and the mean so different?

Can you find 4 numbers that have the same mean and median?

Need: 32 counters for each group of 4 students.

Add "One"

Name _____

 **plore:** If each group member gets the number of counters shown, what is the mean? _____

Your Turn

Case 1:

Suppose one more student joins your group with 9 counters.

- Record the new number of students in the group.
- Share so that each person has the same number of counters.

What is the new mean? _____

Case 2:

Suppose the extra student brings only 4 counters instead of 9.

- Share again and find this mean: _____

 What happened to the mean in each case? Why?

 What would have to happen for the mean to decrease?

Need: 25 counters for each group of 4 students.

Data, Chance and Probability Activity Book, 6–8
© 1994 Learning Resources, Inc.

Could it Be?

Name_____

 **Explore:** In each group of 4, 1 member gets 3 paper squares and the other members get 2 paper squares apiece. Estimate the mean.

Your Turn

- Record the number of people in your group.
- Each person takes paper squares to match a number shown above.
- Record the number of paper squares each person took.
- Record the total number of squares for the group.
- Share so that each person has the same number of paper squares.

How many paper squares does each person have now? _____

 Describe how your group found the mean.

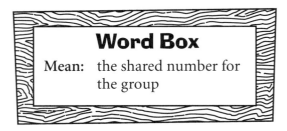

Word Box

Mean: the shared number for the group

 Add a fourth number to this set so that the new mean is 2.5.

Need: 10 paper squares for each group of 4 students.

What's the Total?

Name _____

 **plore:** What is the total number of counters so the mean for your group is 3? _____

Your Turn

- Record the number of people in your group.
- Predict the total number of counters.
- Share this number of counters with group members.
- Find the mean. Is it 3? (If not, try again.)

 How can you find the total if you know the mean and the number of people in the group?

 What is the total number of counters for the group if the mean is 4?

Need: 32 counters for each group of 4 students.

Data, Chance and Probability Activity Book, 6–8
© 1994 Learning Resources, Inc.

Mean, Median, and Mode

Name

 **plore:** How many counters will each of 5 people get if 4 is the mean, median, and mode?

Your Turn

- Work in a group of 5. Show the number each could get.
  (Hint: Look back at what you did for *What's the Total?*

Describe how you found:

- The mean.

- The median.

- The mode.

Can you find a different set of data to fit the problem?

Need: 25 counters for each group of 5 students, "Write About It" from *What's The Total*, page 47.

How Many Are Missing?

Name_____

 **Explore:** How many counters should a fourth student have to change the mean from 4 to 5?

> **Given:** Mean = 4 counters for group of three students

Your Turn:

- Work together.
- What is the total number of counters for the first 3 students? _____
- What is the total number of counters for all 4 students? _____
- Show the number the fourth student should have: _____

 Describe how you found the number of counters the fourth student should have to change the mean.

 How many counters could two students bring to the group of three to change the mean from 4 to 5?

> **Need:** 25 counters for each group of 4 students.

Teaching Notes
Different Ways

Warm-Up

This section focuses on the number of ways to arrange or select a set of objects. To introduce this section, ask the class to predict how many different pizzas can be bought at a favorite pizza place. Discuss the sizes of pizza and the different kinds of toppings and crusts.

Using the Pages

Pizza Today? *(page 51):* Provide more counters than are needed, and monitor the way students set up the chips. This activity involves the multiplication principle. The number of different kinds of pizza available from the menu equals 3 (sizes) x 2 (toppings), or 6 kinds of pizza. If both toppings are ordered on one pizza, then there are 7 kinds.

Two from the Deck *(page 52):* This activity explores the possible outcomes of drawing 2 colors of cards. In the case of two cards, the number of possible outcomes is as follows:

   2 (colors possible for the first card) x 2 (colors possible for the second card) = 4 possible outcomes.
In the case of three cards, the number of possible outcomes is 8. The general rule for the number of possible outcomes for *n* cards is 2^n.

Two out of Four *(page 53):* In this activity the cards are drawn but not replaced. Each of the four cards has an equally likely chance of being drawn first, and each of the remaining three cards has an equally likely chance of being drawn second. In all there are 12 possible outcomes.

Wrap-Up

Ask the class to consider the number of arrangements of three different subjects they study every day. To model the situation, ask three students to stand at the front of the room and hold three different textbooks. Then have students arrange themselves differently while the class records each arrangement. Students should notice that the number of ways is 3 x 2 x 1 for three subjects. Next, have students repeat the activity for four subjects, and note that the pattern continues, that is, 4 x 3 x 2 x 1.

Pizza Today

Name _____

 **Explore:** How many different kinds of pizza can you order?

Your Turn

- Use red chips(R) for pepperoni, white chips(W) for cheese.
- Mark the chips S(small), M (medium) and L(large). Show the different kinds of pizza you can order. For example, SW is "small cheese." Tally them.

 How many different combinations or kinds of pizza did you find?

 Suppose the menu included the choice of thick or thin crust. How many different kinds of pizza can you order?

Need: 12 red-white chips and wipe-off markers.

Two from the Deck

Name _____

Explore: One card is drawn, the color is noted, and the card is replaced. A second card is drawn.

How many different color arrangements can there be?

Your Turn:

- Use the deck of cards. List the color arrangements when 2 cards can be drawn.
- Tally them.

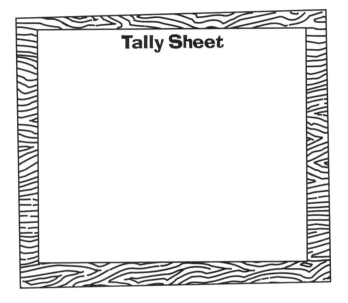

Tally Sheet

How many different arrangements did you find?

How many different color arrangements would there be if 3 cards were drawn?

What is the pattern for finding the number of color arrangements?

Need: Deck of 52 cards.

Two Out of Four

Name

 **plore:** Two cards are drawn, one after the other.

How many different ways could this happen?

Your Turn

- Place the four cards shown in "Explore" face up.
- Show the different ways you can draw 2 cards.
- Tally them.

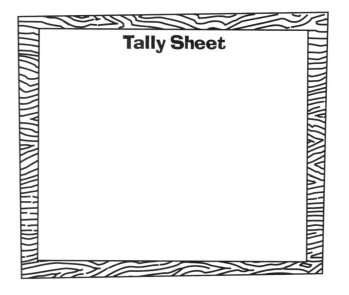

Tally Sheet

How many different outcomes did you find?

How many different outcomes would there be if 3 cards were drawn?

How does order make a difference in counting the outcomes?

Need: Four spade cards shown above.

Data, Chance and Probability Activity Book, 6–8
© 1994 Learning Resources, Inc.

Teaching Notes
Now Sports

Warm-Up

In a school basketball game, the coach has an option of sending one of two players to the line for a "1 and 1" situation. Which one will the coach use: a 70% shooter or a 65% shooter?

Present the above sample question, and discuss the following:
- How were the shooting statistics determined?
- How would they be used by the coach in decision making?
- What other factors or statistics might the coach take into account? For example, the coach might know that the 65% shooter is better in a high-pressure situation.

For the activities in this section, glue spinners to tagboard or file folders. Cut out and laminate them, so they can be marked with wipe-off markers.

The activities in this section involve simulation. Help students realize that data from real world situations are used to construct probability models. Remind students that when simulations are carried out, a small number of trials may produce extreme results. Consequently, they need to combine data to obtain a more reliable result for the probabilities.

Using the Pages

A 50% Shooter *(page 56):* Be sure students mark the appropriate spinner with two outcomes – basket and no basket. For the 50% shooter, the probability of making a basket is 1/2, and the probability of making no basket is 1/2. Similarly, in "Tell and Share," students should notice that the probability of Ricardo's making a basket is 5/12, and the probability of missing is 7/12.

Two Free Throws *(page 57):* Students determine Nicholas' chances of making two successive shots. One trial in the simulation will be two spins of the spinner, and students record whether "Both" or "Not Both" shots were made.

To make a class graph, students can record data in tallies of 5 or 10 on small paper squares. Note that the theoretical probability of making both free throws is 1/4, since there are four equally likely outcomes (BB, BM, MB, and MM), where B means "Ball" and M means "Miss." Only one of these outcomes (BB) is favorable.

A 30% Average *(page 58):* The probability of Florentia's making a hit is 3/10. Use a spinner with 10 outcomes, with 3 sectors marked "Hit" and 7 sectors marked "No Hit." Students need to spin *three* times to represent Florentia's three bats. After those three spins, students should tally either "All 3" or "Not All 3."

It may help students to dramatize the situation by actually imagining they are batting when they spin the spinner. This will reinforce the notion that they must spin three times before recording a tally. In "Write About It," Florentia's simulated value should be close to the theoretical value of 0.027. In "Tell and Share," the theoretical probability of Monica's making three consecutive hits is 1/5 x 1/5 x 1/5, or 1/125 (0.008). Students are not expected to give this result, but they should at least notice that Monica's probability will be smaller than the simulated value for Florentia.

Steffi's Serve *(page 59):* If necessary, discuss serving in tennis so that students are aware of the meaning of the terms "Fault" and "Double Fault." In this simulation, students are expected to fill in missing pieces. For example, they need to insert labels on the tally box and spin the spinner *2 time*s for each trial.

In "Tell and Share," students should note that Steffi's chance of serving a double fault on a bad day, namely, 0.6 x 0.6 or 0.36, will be considerably more than her chance of serving a double fault as simulated in the experiment, which would be about 0.2 x 0.2, or 0.04.

Up to Bat *(page 60):* Batting averages are typically reported to three decimal places. To simplify the simulations, batting averages are rounded to two places.

In this activity, students are expected to complete some aspects of the simulation. For example, spin *3 times* and label the Tally Sheet categories, "2 or 3 Hits," "1 Hit," or something equivalent. In "Tell and Share," students need to list only 8 outcomes: HHH, HHN, HNH, HNN, NHH, NHN, NNH, and NNN, where H means "Hit" and N means "No Hit." Some students may list 27 outcomes, using hits, walks, and outs. This is a correct interpretation of the problem, but more complicated than necessary.

One of Each *(page 61):* In this activity the students provide most of the information for the simulation. To tally the situation, they need to label the Tally Sheet "One of Each," "Not One of Each," or something equivalent. The simulated value should be close to the theoretical probability of 0.11. In "Tell and Share," there are 27 possible outcomes; 6 of these are 1 hit, 1 walk, and 1 out.

Stefan's Smashes *(page 62):* If necessary, discuss "smashing" in tennis. Quite often, a smash is a hit for a winner by top players. Stefan actually smashes for a winner 9 times in 10. This means that 1 time in 10, he makes an error or the other player returns the smash.

In "Tell and Share," the theoretical probability of "smashing all four" is 0.66, so the simulated value should approximate this amount. Since the probability of "Smashing All Four" is 0.66, the probability of "Not Smashing All Four" covers all the remaining possibilities at 1 - 0.66, or 0.34.

Wrap-Up

Team up students and ask them to pose sports theme problems whose solutions require a simulation. Then have teams exchange problems. Each group needs to follow these steps:

- Collect data for the problem.
- Express the data as probabilities.
- Formulate the problem.
- Decide how the problem should be simulated.
- Exchange with another group.

A 50% Shooter

Name_____

 **plore:**

Free Throws

Nicholas: 60 out of 120
Ricardo: 50 out of 120

Who is the 50% shooter?

Your Turn

- Suppose the 50% shooter takes 9 free throws in the next game. About how many baskets should he make?

- Mark the appropriate spinner to simulate the free throws in his next game. Tally your results.

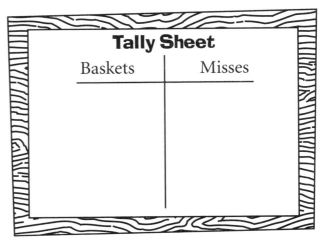

Did the outcome turn out as you predicted?

How does your result compare with the results of other classmates?

If Ricardo had 9 free throws in the next game, about how many baskets should he make?

Need: Activity Master 2.

Two Free Throws

Name _____

 **Explore:** Nicholas is on the free throw line for 2 shots.

What do you estimate his chances are of making both shots?

Free Throws:	
Nicholas: 60 out of 120	

On the scale, mark an X to show the chances of his making both shots.

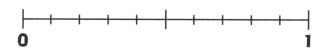

Your Turn

- Find a partner.
- Mark and use the appropriate spinner for the experiment.

The Experiment

- Spin 2 times to simulate 2 free throws.
- Did he make both or not? Tally to show.
- Repeat 10 times, then partner takes turn.
- Make a class graph to show the numbers for "Both" and "Not Both."

Tally Sheet

Both	Not Both

 Did the probability for "both" turn out as you predicted?

 When Nicholas has two free throws, how many outcomes are possible?

How many of these outcomes result in 2 baskets?

Need: Activity Master 2.

Data, Chance and Probability Activity Book, 6–8
© 1994 Learning Resources, Inc.

A 30 % Average

Name _____

 **Explore:** Florentia will bat 3 times in the next game.

Hits
Monica: 4 for 20
Florentia: 6 for 20

What do you estimate her chances are of making 3 consecutive hits?

On the scale, mark an X to show the chances of making 3 consecutive hits.

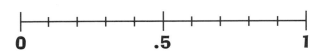

Your Turn

- Find a partner.
- Mark and use the appropriate spinner to simulate Florentia at bat.

The Experiment

- Spin 3 times to simulate 3 hits.
- Did she make all 3 or not? Tally to show.
- Repeat 10 times.
- Partner takes a turn.
- Make a class graph to show the numbers for "All 3" or "Not All 3."

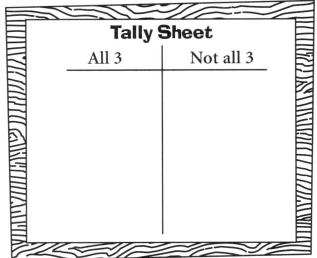

 Draw a scale to show the probability of 3 hits, based on the class data.

How did your data compare with the class data?

 When Monica steps up to bat, what are her chances of making 3 consecutive hits?

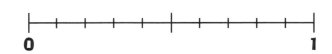

Need: Activity Master 2.

Steffi's Serve

Name_____

 **plore:** Steffi is ready to serve.

What do you estimate her chances are of a double fault?

On the scale, mark an X to show the chances of making a double fault.

Faults	
Steffi: 2 in 10	

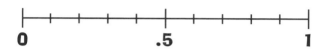

Your Turn

- Find a partner.
- Mark an appropriate spinner to simulate Steffi serving.
- Complete the Experiment and Tally Sheet.

The Experiment

- Spin _____ times to simulate 2 serves.
- Did she make a double fault or not? Tally to show.
- Repeat 10 times, then partner takes a turn.
- Make a class graph to show the numbers.

Tally Sheet

Double Fault	No Double Fault

Based on your class graph, about how often will Steffi serve a double fault?

Was your prediction about right?

On a bad day, Steffi serves 6 faults in 10. How does a bad day affect her chances of serving a double fault?

Need: Activity Master 2.

Data, Chance and Probability Activity Book, 6–8
© 1994 Learning Resources, Inc.

Up to Bat

Name _____

 **Explore:** Lou will bat 3 times in the next game.

What do you estimate his chances are of making at least 2 hits?

Lou's Batting Statistics
Hits: .30
Walks: .10   Outs: .60

On the scale, mark an X to show the chances of making at least 2 hits.

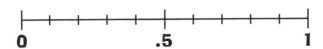

Your Turn

- Find a partner.
- Mark and use the appropriate spinner to simulate Lou at bat. Complete the Experiment and Tally Sheet.

The Experiment

- Spin _____ times to simulate 3 "at bats."
- Did he make _____ or not? Tally to show.
- Repeat 10 times, then partner takes a turn.
- Make a class graph to show the numbers.

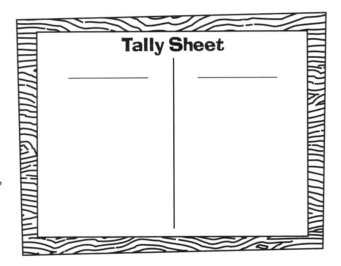

Tally Sheet

Based on the class graph, about how often will Lou make at least 2 hits out of 3 times at bat?

When Lou bats 3 times, how many outcomes are possible?

How many of these include at least 2 hits for Lou?

Need: Activity Master 2.

One of Each

Name _____

 **Explore:** Lou will bat 3 times in the next game.

What do you estimate his chances are of making 1 hit, 1 walk, and 1 out?

Lou's Batting Statistics
Hits: .30
Walks: .10   Outs: .60

On the scale, mark an X to show the chances of 1 hit, 1 walk, and 1 out.

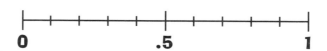

Your Turn

- Find a partner.
- Mark and use the appropriate spinner to simulate Lou at bat.
  Complete the Experiment and Tally Sheet.

The Experiment

- Spin 3 times to simulate 3 "at bats".
- Did he get one of each or not? Tally to show.
- Repeat 10 times, then partner takes a turn.
- Make a class graph to show the numbers.

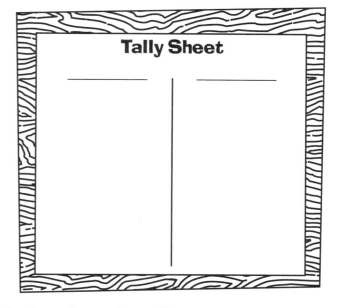

Tally Sheet

 Based on the class graph, about how often will Lou get one hit, one walk and one out in 3 times at bat?

 When Lou bats 3 times, how many outcomes are possible?

How many of these include one of each?

Need: Activity Master 2.

Data, Chance and Probability Activity Book, 6–8
© 1994 Learning Resources, Inc.

Stefan's Smashes

Name _____

 **Explore:** Stefan hits 4 smashes in a set.

What do you estimate his chances are that all four are winners?

Smashes

Stefan: 9 winners in 10

On the scale, mark an X to show the chances of making 4 winning smashes.

Your Turn

- Find a partner.
- Mark and use the appropriate spinner to simulate Stefan's smashes. Complete the Experiment and Tally Sheet.

The Experiment

- Spin each spinner. Tally the outcome.
- Repeat 10 times.
- Make a class graph to show the numbers.
- Partner takes turn.

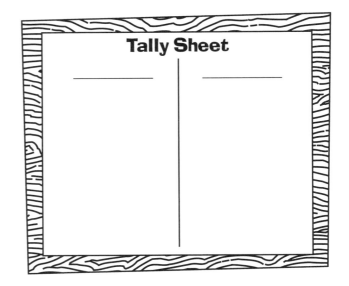

Tally Sheet

 How did your experiment simulate Stefan's smashes?

 Based on the class data, what is Stefan's chance of smashing all 4 for winners? Not smashing all 4 for winners?

Need: Activity Master 2.

Teaching Notes
For the Record

Warm-Up

U-Beat magazine claimed that 60% of middle school students preferred *salty* snacks (popcorn, chips, pretzels) to *sweet* snacks (different candies). Assign students to work in small groups to decide how they would carry out a poll in their school to test this claim.

The focus of this activity is to have students consider sampling a population and interpreting the results of a poll.

Using the Pages

Movie Poll *(page 64):* The question in "Explore" cannot be answered because no information is given about how many students saw both. The extra information can be determined in the simulation using data from the completed table. In "Tell and Share," the simulation models the information given in the community data. However, students cannot compare the simulation with the community data because the community data does not provide information about those who saw both movies.

What are the Chances? *(page 65):* In this activity, students combine the data from the simulation into a Class Data Table.

Read My Mind *(page 66):* A person with no prior knowledge (no mind-reading capabilities) could expect to be right about the card colors about 50% of the time. So a true mind-reader should be able to perform at a level above 50%!

Timed Experiment *(page 67):* Have the class construct a back-to-back stem-and-leaf plot for "portrait" and "non-portrait" groups. In "Tell and Share," students will choose the statistic based on the data. In most situations, the median and the range probably work best.

Box the Timed Data *(page 68):* Try to do this activity on the same day as *Timed Experiment*, if possible. Answers to the "Tell and Share" question will vary, but students should realize that outliers are removed a long way from the central data. The usual criteria are 1-1/2 box lengths above the upper quartile and 1-1/2 box lengths below the lower quartile.

Wrap-Up

Have students repeat *Timed Experiment* and *Box the Timed Data.* Ask the following question: "Would the plots be any different if the roles for each pair were reversed?" Then ask students and their partners to work together to answer this question: "What does repeating the experiment tell you about sampling a population?"

Data, Chance and Probability Activity Book, 6–8
© 1994 Learning Resources, Inc.

Movie Poll

Name

 **Explore:** Half of the students in your community have seen Movie A; 1/3 have seen Movie B. About how many have seen both?

Simulation

- Find a partner.
- Mark two spinners to model the viewing of Movie A and Movie B.
- Use the spinners to simulate interviewing 20 students in one class.

The Experiment

- Spin each spinner. Tally the outcome.
- Repeat until there are 20 tallies.

Tally Sheet

Movie	Seen	Not Seen	Total
A			
B			

 Describe the results of the simulation.

What percent of the 20 students saw both movies?

 Why does the simulation model the movie viewing?

Is the class result different from that of the community?

Need: Activity Master 2 and paper clip for spinners. **Save:** *Movie Poll* for page 65.

What are the Chances?

Name _____

Explore: Half of the students in your community have seen Movie A; 1/3 have seen Movie B.

What is the probability that a student chosen at random has seen both?

Class Activity

- Work as a class.
- Add your data from *Movie Poll* to the Class Data Table.
- Work together to keep a running total on the data.

Tally Sheet

Movie	Seen	Not Seen	Total
A			
B			

How many students were polled? If one student's name were drawn from a box, what is the probability that this student saw:

- both movies?

- one movie?

- no movies?

If you know that a student has seen Movie A, what is the probability that the student has seen Movie B?

Need: Data from *Movie Poll,* page 64.

Read My Mind

Name_____

 **Explore:** Your partner draws one card and says, "Read my mind and tell me what color it is."

Would a mind reader be right more than 50% of the time?

Your Turn

- Find a partner, and do the Experiment.

The Experiment

- Shuffle the deck. Draw one card, but don't tell your partner what it is.
- Partner tries to "read your mind" to tell the color of the card.
- Tally whether partner is correct or not correct.
- Repeat 20 times.
- Partner takes a turn.

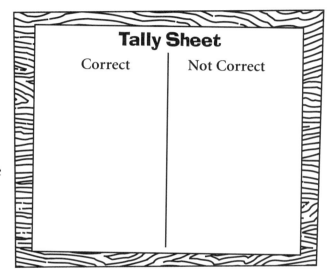

 What was the probability that your partner guessed the card right? That you guessed it right? Compare.

 Why should a true mind reader be able to guess it right more than 50% of the time?

Need: Deck of cards.

Timed Experiment

Name_____

 **Explore:** Would a student drawing a self-portrait or a student doing nothing estimate a minute better?

Your Turn

- Find a partner.
- Toss a chip to decide which student begins to draw a self-portrait.
- Do the Experiment.

The Experiment

- Note the time, and tell your partner to estimate one minute.
- Note the time when your partner says "one minute." Record the elapsed time on the Tally Sheet.
- Switch roles.
- Make a class back-to-back stem-and-leaf plot.

Tally Sheet

Portrait	No Portrait
sec.	sec.

How do the groups compare?

Was your estimate correct?

What statistical term could you use to compare the two groups (mean, median, mode, range, lowest/highest score)?

Need: Watch or clock with a second hand and 1 red/white chip.
Save: *Timed Experiment* for page 68.

Data, Chance and Probability Activity Book, 6–8
© 1994 Learning Resources, Inc.

Box the Timed Data

Name

 **Explore:** Using the Timed Experiment plot, which group's estimates were the least scattered?

Class Activity

- Work as a class.
- Use the back-to-back stem-and-leaf plot from *Timed Experiment* to make a back-to-back box-and-whisker plot.
- Mark any outliers with an ✶.

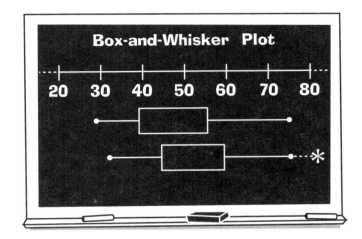

 How do the times of the groups compare?

Which group had more scattered times?

Word Box
Outlier: extreme value

 How can you determine an outlier more precisely?

Need: Back-to-back stem-and-leaf plots from *Timed Experiment*, page 67.

Teaching Notes
Chips

Warm-Up

The activities in this section focus on simulating the number of peanut butter chips in cookies using random objects, such as spinners and cards. Ask students what they would use to simulate the chance of the last digit of someone's house number being even. In this situation, since there is a 50-50 chance of odd-even, tossing a fair coin (equal chance of heads or tails) can be used.

Using the Pages

Peanut Butter Chip Cookies *(page 70):* Students need to consider why a spinner with eight equal-sized sections is a suitable model for randomly distributing the peanut butter chips. Each of the eight cookies corresponds to one of the sections on the spinner. To obtain a good estimate, have students combine their results from the experiment. The estimate should be around 0.93.

No Chips *(page 71):* The simulation is almost identical to the previous activity except that cards are used instead. The estimate should be around 0.14.

Wrap Up

Encourage students to formulate similar real problems, and determine ways of simulating them. For example, use weather statistics involving some given chance of rain during a particular month.

Data, Chance and Probability Activity Book, 6–8
© 1994 Learning Resources, Inc.

Peanut Butter Chip Cookies

Name _____

 **Explore:**

A mix for 8 cookies has 20 peanut butter chips.

What is the probability that each cookie has at least one peanut butter chip?

Your Turn

- Find a partner.
- Number the appropriate spinner 1 to 8 to match the 8 cookies.
  Use the spinner in the Experiment.

The Experiment:

- Spin to see which cookie gets the chip. Tally.
- Repeat 20 times.

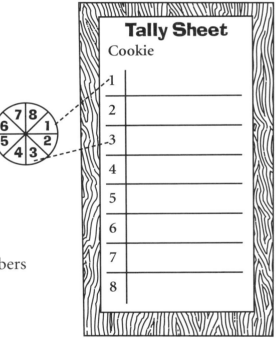

What did the 8 spinner numbers model?

Why did you spin 20 times?

Based on your data, how many cookies got at least one peanut butter chip?

What is the probability that a cookie has at least one peanut butter chip?

Need: Activity Master 2.

No Chips!

Name _____

**plore:** A mix for 26 cookies has 50 peanut butter chips.

What is the probability of getting a cookie with no peanut butter chips?

Your Turn

- Find a partner.
- Use the 26 red cards to model the 26 cookies in the experiment.

The Experiment

- Shuffle the red cards.
- Draw a card, tally, replace the card, and shuffle the deck.
- Repeat 50 times.

Tally Box

	A	2	3	4	5	6	7	8	9	10	J	Q	K
♦													
♥													

What does each tally mark represent?

How many cookies have no peanut butter chips?

What is the probability of getting no peanut butter chips?

Need: Deck of cards.

Data, Chance and Probability Activity Book, 6–8
© 1994 Learning Resources, Inc.

Activity Master 1
Target Toss

2

5

2

Activity Master 2
Spinners

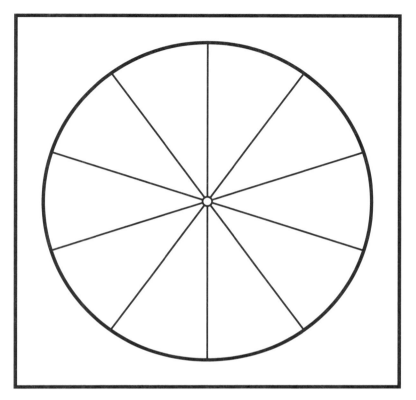

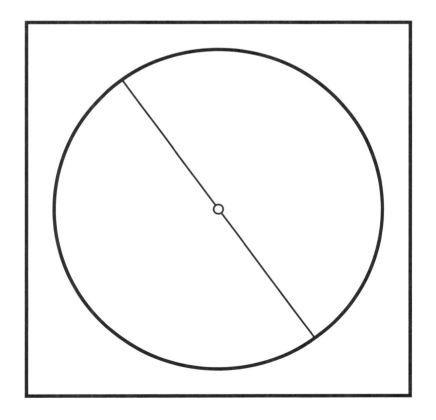

Data, Chance and Probability Activity Book, 6–8
© 1994 Learning Resources, Inc.

Activity Master 2 (Continued)
Spinners

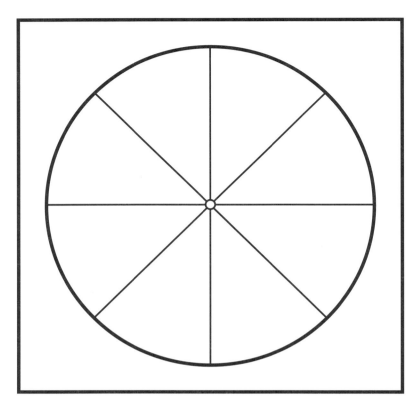

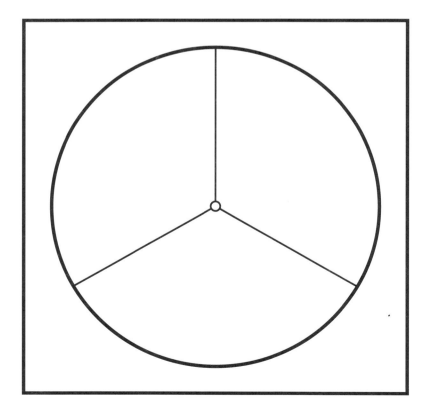

Activity Master 3
Rollers

Activity Master 4
Cups

Cup 1:

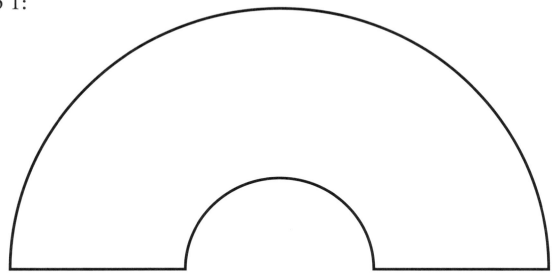

Cup 2:

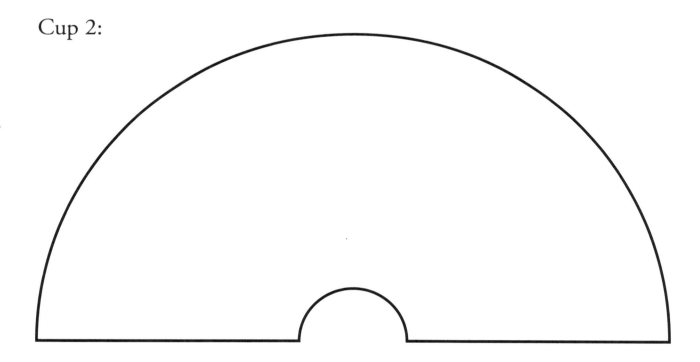

Progress Chart

Name _____

Grade _____

Date Started _____

Date Finished _____

On Target
- ☐ Target Toss
- ☐ Line Them Up
- ☐ Plot It Again
- ☐ Back-to-Back

Deal the Deck
- ☐ Black for Sure?
- ☐ Equally Likely
- ☐ Why Equally Likely?
- ☐ Black More Likely!
- ☐ Why Black?
- ☐ Match the Cards
- ☐ Even Things Up
- ☐ The Packer's Deck
- ☐ Some Hearts
- ☐ Which Two?
- ☐ Double Red

Movie Time
- ☐ How Many Movies?
- ☐ Back-to-Back
- ☐ Quartiles
- ☐ Boxing the Movie Data
- ☐ Line Plot for the Movie Data
- ☐ Box Plot Scale
- ☐ Favorite Movie
- ☐ Plot the Rank Order
- ☐ Back-to-Back Line Plot

Cups and Rollers
- ☐ Roll the Roller
- ☐ Cup Ups and Downs

Averages
- ☐ Finding the Average
- ☐ Zero In or Out?
- ☐ Locating the Mean
- ☐ Mean or Median?
- ☐ Median and Mean
- ☐ Add "One"
- ☐ Could It Be?
- ☐ What's the Total?
- ☐ Mean, Median, and Mode
- ☐ How Many Are Missing?

Different Ways
- ☐ Pizza Today
- ☐ Two from the Deck
- ☐ Two Out of Four

Now Sports
- ☐ A 50% Shooter
- ☐ Two Free Throws
- ☐ A 30% Average
- ☐ Steffi's Serve
- ☐ Up to Bat
- ☐ One of Each
- ☐ Stefan's Smashes

For the Record
- ☐ Movie Poll
- ☐ What are the Chances?
- ☐ Read My Mind
- ☐ Timed Experiment
- ☐ Box the Timed Data

Chips
- ☐ Peanut Butter Chip Cookies
- ☐ No Chips

Data, Chance and Probability Activity Book, 6–8
© 1994 Learning Resources, Inc.